COVID-19 and Beyond: Managing the Unexpected

Held at the 2021 AIChE Annual Meeting

Boston, Massachusetts, USA and Online
7 - 11 November and 15 - 19 November 2021

ISBN: 978-1-7138-5684-9

COVID-19 and Beyond: Managing the Unexpected

Held at the 2021 AIChE Annual Meeting

Boston, Massachusetts, USA and Online
7 - 11 November and 15 - 19 November 2021

AIChE
The Global Home of Chemical Engineers

Printed from e-media with permission by:

Curran Associates, Inc.
57 Morehouse Lane
Red Hook, NY 12571

Some format issues inherent in the e-media version may also appear in this print version.

AIChE
120 Wall Street, FL 23
New York, NY 10005-4020

Phone: (800) 242-4363
Fax: (203) 775-5177

www.aiche.org

Additional copies of this publication are available from:

Curran Associates, Inc.
57 Morehouse Lane
Red Hook, NY 12571 USA
Phone: 845-758-0400
Fax: 845-758-2633
Email: curran@proceedings.com
Web: www.proceedings.com

TABLE OF CONTENTS

Teaching Process Calculations to Undergraduate Chemical Engineers: An Engaged & Energized Online Class 1
Preeti Aghalayam, Kartic Vaidyanathan

From MOOCs to Residential Instruction — Departmental Perspective on Digital Learning @ MIT ChE 4
Zongyu Gu, Martin Z. Bazant

Virtual Implementation of a Hands-On Learning Tool and Its Effect on Student Comprehension and Motivational Gains 5
Kitana Kaiphanliam, Olivia Reynolds, Aminul Islam Khan, Olufunso Oje, David B. Thiessen, Prashanta Dutta, Olusola Adesope, Jacqueline Burgher Gartner, Bernard Van Wie

Developing Simulation-Based, Student-Centred Educational Software in Process-Control Education and Training 6
Niall English

Reinventing Unit Operations Laboratories Post-COVID-19 7
Samira Azarin, Christopher Barr, Janie Brennan, Tracy Carter, Amy J. Karlsson, Sarah Wilson

Replacing Exams with Projects: Advantages and Disadvantages as Observed During Hybrid and Remote Learning and Broader Efforts Towards Equitable Assessment 8
Lucas Landherr

Lessons Learned During the COVID-19 Pandemic: A Literature Synthesis 9
Deja Preusser, Milo Koretsky

Interventions to Mitigate Loss of Community During COVID-19 10
Christi Patton Luks

Oral Exams in Undergraduate Cheme Thermo Taught Online 12
Gennady Gor

Online Cheating: A Darker Side of Educational Technology 13
Javier Huayta, Zachary S. Campbell, Matthew Cooper

Aiche Education Division Virtual Communities of Practice (VCPs): Updates from the 2020-2021 Academic Year 14
Daniel Lepek, Matthew Liberatore

Modeling the Progression of Fibrosis with Dysregulation of ACE2 in COVID19 Patients 15
Mohammad Aminul Islam, Ashlee Ford Versypt

DNA-Directed Patterning to Validate a Liposome Model of Sars-Cov-2 16
Molly Kozminsky, Thomas R. Carey, Lydia L. Sohn

Polymersomes Decorated with Sars-Cov-2 Spike Protein Receptor Binding Domain Elicit Robust Humoral and Cellular Immunity 19
Lisa Volpatti, Rachel Wallace, Shijie Cao, Michal Raczy, Ruyi Wang, Melody Swartz, Jeffrey A. Hubbell

Inhalation Delivery of RNA-Loaded Lipid Nanoparticles Against Sars-Cov-2 20
Kai Slaughter, Daniela Isaacs-Bernal, Ursula Nosi, Eric Donders, Chang Xue, Timothy Cheung, Tekeleselassie Woldemariam, Darryl Falzarano, Amy Wong, Brian Cox, Molly S. Shoichet

A Human Pluripotent Stem Cell Approach to COVID-19 (Invited Speaker) 23
Joyce Chen

Application of Biophysical and Chemical Engineering Principles for Understanding Molecular Scale Interactions Critical to Virus Entry and Infection of Its Host (Invited Speaker) 24
Susan Daniel

Commercialization of Aptamer-Based Electrochemical Sensors 25
Edgar D. Goluch

(Invited Talk) Developing Nanosensor Technology to Detect Cancer in Patients 26
Daniel Heller, Zvi Yaari

(Invited Talk) Wearable Health Sensors 27
Chelsea Monty-Bromer

Resilience in the Face of COVID-19: Managing a User Program at the Molecular Foundry 28
Shannon Ciston

Global Supply Chain Optimization for COVID-19 Vaccines Under Covax Facility 32
Apoorva Katragadda, Xiaonan Wang, Iftekhar Karimi

Sustainability and Resiliency Assessment for Industrial Synergies Between Renewable Energy and Chemical Production 34
Apoorva Bademi, William Farlessyost, Miriam Stevens, Shweta Singh

A Community-Level Energy Planning Model for the Dynamic Optimization of Waste-To-Gas and Waste-To-Power Anaerobic Digestion Supply Chains 35
Irene Mas Martin, Alvina Aui, Lisa Schulte, Mark Mba Wright

Asset Integrity Management (AIM) for Hydrogen: Major Incident Risk Reduction by Integrating a Risk-Based Safety Approach 36
Gareth Ellor

More Haste, Less Speed: Proportionate Supply Chain Risk Management for the Accelerating Hydrogen Economy 37
Gareth Ellor

Leading and Empowering Global Teams During a Pandemic 38
Javier Guzman

Navigating the COVID-19 Pandemic at Michigan Technological University 39
Pradeep K. Agrawal, Jeanna Collins, Kurt A. Rickard, Tony N. Rogers

Managing Teams Effectively During a Pandemic Required – Empathy, Connection, Agility, Prioritization and Support 40
Rajat Duggal

Lessons Learned from Field Testing a New Technology During the Pandemic 41
Anthony Ku

Author Index

625791 Teaching Process Calculations to Undergraduate Chemical Engineers: An Engaged & Energized Online Class

Abstract:

Introduction

"Process Calculations" is a required class in undergraduate ChE curricula worldwide. It serves not only as a good introduction to the subject but is also a pre-requisite for many ChE courses including Reaction Engineering, Separation Processes and so on. At IIT Madras, the course is in the first year of the four year undergraduate program, and sees a typical class size of ~100 students. Managing the stress levels of the students, alongside the learning goals for the course and the evaluations as required by administration, is a major challenge. In this talk, we want to illustrate a few best practices, applicable to online/hybrid classes in particular, with examples of student performance and testimonials.

Philosophy & guiding principles

In putting together the details of the course, we wanted to ensure that the first year undergraduate students felt a lot of excitement for chemical engineering. Second, examinations and evaluations were causing unprecedented levels of stress and competitiveness, and we wanted to do better with those, with the constraints of the existing systems. Finally, as the class was completely online, and the usual campus environment was absent, we wanted to shoulder an additional responsibility of building a sense of belonging to IIT Madras.

We are proponents of what we call "Play to learn" or "game based learning" at IIT Madras. Over the past few years, both in physical classrooms and in online ones, it has been our attempt to bring in more play and fun. In an elective course titled "Let us play to learn" we have students custom-building games and interactivities – these include board, mobile hypercasual, quiz-based & card games – for subsequent use in various classes. The ideas are gaining good popularity in both the student and faculty communities. We wanted to harness our learnings from this work, into a full-fledged core engineering course, and *Process Calculations*, as an introductory ChE core course, provided a great platform for the same.

Overall the goal of this teaching experiment was to provide an engaged and energized learning of the objectives of the course, to a class of 100+ students, in an online mode, using the broad guiding principles of game based learning.

Tools and Techniques

A detailed description of the evolution of the course is provided in this section. While several elements are common in chemical engineering pedagogy, the unique challenges due to completely online instruction, for a population that is not used to it, are addressed with new measures that are easy for instructors and learners to adopt. A few of the unsolved issues are highlighted as well, to enable further discussions, and future improvements.

- **Resource Materials**

The standard textbook used at IIT Madras for this course is the popular 'Elementary principles of chemical processes' by Felder and Rousseau. We found that it was particularly important to stick to the data tables, units, and graphical information in this book, to avoid confusions. In addition, a handful of short videos (10-15mins) long to described via animations, for the more difficult/confusing concepts were provided. All the lecture slides, and video recordings of the classroom interactions were made available to students as well.

- **Interactive Learning**

A common challenge faced by instructors in such large classes is getting students to interact – to ask or answer questions. The online mode further exacerbated this issue. It has been our finding that mixing in "fun" and "serious" questions, particularly via anonymous, online polling, is very helpful. Rating, word cloud, and open text modes of polls were also frequently used in class. A text chat-based discussion forum, worked very well in a live class, as well.

- **Bite Sized Fun Learning & Assessment**

We used short quizzes to provide well-rounding learning, in two ways. One, was a self-assessment tool, built like an online game with some fun graphics (bowling alleys to get full strikes at; mountains to climb, etc). The second was a graded "summary" quiz at the end of every class. Students routinely appreciated these interventions that helped them keep up-to-date on the course material, while having fun in a fail-safe manner.

- **Enabling Peer Learning**

A 30 mins breakout room session, during class, to enable students to discuss the tutorial solution methodology in groups, was provided. Students later individually solved the problems and uploaded their answers for checking by the teaching assistants. Possibly due to the students being in their homes (not on campus), the effectiveness of breakout rooms was not as good as could be.

- **Quick Feedback and In-depth Reflections**

Through the course of the semester, it was critical to not only remain open to receiving feedback from students, but also to act upon it with immediacy. The inputs regarding how the learning and progress were proceeding came up in various ways – a quick and immediate "reaction" (thumbs up/down, smiley/frowny faces) in the online class helped gauge if majority had understand the explanation provided. At the end of every week of classes, students were encouraged to answer open ended questions about the taught material, comment on their own learning and progress through the course, via an online form.

- **Assessments**

Three exams were conducted in the course. One was a rapid fire online quiz (with a leaderboard and some memes for fun), the second was based on typing numerical answers in an online form, and a final exam involving hand written solutions and uploading, were conducted. These exams were graded in the conventional manner, with emphasis on getting the right answer and the correct sequence of steps; and contributed 50% of the final grade, whereas the daily quizzes, class participation, reflections & tutorials contributed the rest. Overall, the goal was to give a judicious mix of points for effort and accuracy, while fitting into the constraints set by the institute administration.

Analysis

A few important motivating factors drive us to recommend these as good – if not best - practices to others. First, the level of interaction – in a typical in-person class, 10-20 students at the maximum answer or ask questions. The roster and specific calling out of students used here, quickly moved this number up to ~95 students (out of 118 total registered). Second, the reflections forms proved to be a great way to obtain in-depth feedback from students. Many of the comments could be used directly to improve the course experience immediately, in the next week.

Students frequently expressed their relief at being able to learn about and build on chemical engineering concepts, in a less judgmental, uncompetitive, fun manner. Balancing effort with quality, with tools used so students could self-evaluate (via games like the bowling alley one), were definitely ideas that bore rich dividends. The polls, were helpful to gauge the mood of the class, and served as good de-stressors as well.

Some of the tough parts of a Process Calculations course involve the use of diagrams – like the Psychrometric Chart in design of air-conditioning systems. While the use of the basic principles of mass and energy balances are well illustrated in such applications, they have often been a source of fear and frustration among undergraduate ChE students. With the game-based learning and very interactive mode of instruction used here, several students voluntarily stated that this was their "favourite" topic in the course!

Some of the final comments by students are shown below:

- *Well, I was not a very big fan of chemical reactions and inorganic chemistry. Earlier thought that chemical engineering=chemist. After this course, I realised that chemical engineering is just the perfect branch for me..*
- *This course taught a lot of practical things, and all these concepts and problem solving skills will really be very helpful to us as chemical engineers. The course structure was organized and that made it interesting to study it.*
- *The best part was interactive session and time to time quiz for regular evaluation.*
- *CH is the only course that doesn't feel disjoint from reality*

Some of the elements of the course that worked less well were – peer interaction and tutorials. In the initial part of the semester, the students were reluctant to speak in their breakout rooms. Scanning and uploading hand-written answers was also a challenge for those with poor internet connectivity/older devices. Light weight apps or online modules integrated into learning management systems are probably better in terms of uniformity of learner experience.

Conclusions

The Covid pandemic has led to a major upheaval in our teaching-learning. Forging friendships, co-operative learning, and development of professor-student bonds have proved particularly difficult in these times. Using the principles of game-based learning in the first ChE course of their curriculum, can be critical in retaining their interest and enthusiasm for the subject. The evaluations also indicate an upswing in the class average marks, underscoring the possibility of integrating the prescribed grading methodologies, within these new ideas. The various online tools opened the doors to several new means for energizing and engaging students, and it is our hope that these will find a regular place in the future.

627906 From MOOCs to Residential Instruction — Departmental Perspective on Digital Learning @ MIT ChE

Abstract:

The Department of Chemical Engineering at MIT has seen a rapid growth in digital learning activities over the past few years. In this talk, we will showcase completed and ongoing digital learning projects and discuss the department's strategic vision. We will present three Massive Open Online Courses (MOOCs) on edX: "10.50.1x Analysis of Transport Phenomena I: Mathematical Methods", "10.50.2x Analysis of Transport Phenomena II: Applications", and "10.S95x Physics of COVID-19 Transmission"; one Small Private Online Course (SPOC): "10.MBCx Math Boot Camp for Engineers"; and other residential projects using digital tools guided by research-based teaching practices. We will share the lessons learned from producing digital educational contents, including videos, graphics, and assessments. Finally, we will explain our departmental strategy for continuing our digital learning work both residentially and globally.

629036 Virtual Implementation of a Hands-on Learning Tool and Its Effect on Student Comprehension and Motivational Gains

Abstract:

Our project involves the national dissemination of highly visual hands-on learning tools focused on fluid mechanics and heat transfer principles to 44 institutions and branch campuses within the United States. Like many other educators, our team had to adapt the implementation protocols to accommodate remote learning during the COVID-19 pandemic. Rather than students working in groups with our hands-on learning tools, we created follow-along video implementations and supplementary tutorial videos. The videos allow students to complete the complementary worksheets associated with each hands-on learning tool while watching a graduate student explain basic concepts and collect real-time data with the hands-on learning tools. The supplementary tutorial videos are focused on an in-depth discussion of a single conceptual aspect of the learning tool. Across three remote-learning semesters, a total of 36 virtual implementations at 12 institutions were completed with approximately 630 chemical and mechanical engineering students. An asynchronous implementation method was used for 70% of the virtual implementations, while other instructors presented virtual material in a synchronous virtual setting with several allowing live, small-group discussion. At the conference, we will present conceptual and motivational assessment results to compare the effectiveness of virtual implementations versus traditional interactive hands-on implementations.

623434 Developing Simulation-Based, Student-Centred Educational Software in Process-Control Education and Training

Abstract:

Laboratory experiments demonstrating important underlying principles of process control are often challenging in their capacity for scalability towards larger-numbers sessions for students, and are often very time- and resource-consuming. Indeed, the recent rush to embrace online teaching and learning in view of the COVID-19 pandemic has accelerated very substantially since Spring 2020 an already-emerging theme of software-based exploration of key concepts in process control for students by means of virtual experiments.

The present contribution has as its central hypothesis and goal the ascertaining of how the complete implementation of virtual-laboratory environments (VLE) can effect positive learning outcomes of a student via constructivist pedagogical approaches. In so doing, a modularised MATLAB-and-Simulink VLE platform has been developed, accompanied by appropriate web-tools, which is focussed specifically on this constructivist approach to student-centred learning. To assess the effectiveness of this VLE, volunteers were split into two groups - the reference and VLE group, both of which took part in the same temperature-control exercise (via PID control). The reference group were exposed to the more traditional behaviouristic practical mode of learning, where they would pay close attention to the actual real-world experimental system and results at UCD Chemical Engineering laboratories; here, the material is gone through in a more prescribed manner, where the student has a certain level of engagement in their own learning, as per classical theory. In contrast, the VLE group encountered the equivalent specifically-designed virtual environment, where they worked in groups and individually to engage with a virtual version of the PID temperature-control experiment, based on the real-world system, where they can vary input parameters and controller settings and explore themselves the system response in time and Laplace domains, with trial-and-error approaches lending themselves to determining optimal settings.

Using equal time allocations to both approaches, it was set up - by design - that the only possible factor between the outcomes for both groups was the pedagogical approaches taken to training. Upon completion, quick MCQ examinations were undertaken to assess the groups' learning outcomes, and it was found that the VLE cohort displayed a generally superior level of understanding; the level of active self-engagement with the topic was found to correlate reasonably well with the degree of understanding. This suggests that VLEs, designed carefully and done well with appropriate software examples, can lead to positive and effective training experiences.

621678 Reinventing Unit Operations Laboratories Post-COVID-19

Abstract:

With the emergence of COVID-19 in Spring 2020, faculty were forced to quickly convert face-to-face courses to online offerings. While this transition was a challenge across all courses, the need to move a hands-on laboratory experience to fully online required significant course modifications. As a result, faculty were forced to focus on laboratory learning outcomes to ensure that student learning goals were met in this non-traditional setting. Surprisingly, this opportunity to reflect on learning outcomes allowed many of the weaknesses of "traditional" unit operations courses to be identified. We will present our assessment of chemical engineering laboratory learning outcomes and how they are currently being satisfied in a traditional face-to-face unit operations course. To better meet the laboratory learning outcomes moving forward, we will discuss strategies for modernizing the teamwork, communication and problem-solving skills being taught in the lab through reinventing the unit operations course curriculum. Additionally, we will discuss two potential surveys: one to probe a broader set of unit operations instructors, and another to probe industry professionals. These surveys will ensure that curriculum changes meet both the standards and needs of academics as well as the ever-changing needs of engineers in industry. With this work, we aim to modernize the unit operations curriculum to better prepare students for their careers as working engineers.

631200 Replacing Exams with Projects: Advantages and Disadvantages As Observed during Hybrid and Remote Learning and Broader Efforts Towards Equitable Assessment

Abstract:

In response to a transition to remote and hybrid learning from the COVID-19 pandemic, as well as an environment with generally higher levels of stress that can cause students difficulty to prove their knowledge and ability, many instructors have tried to find ways of changing their means of evaluating student performance and understanding. The instructors' efforts have also been impacted on searching for methods that are more equitable to students of all backgrounds and experiences, and so the assessment reflects less inherent bias. Some methods have included weekly quizzes as opposed to larger exams, oral discussion exams instead of written tests, and replacing exams with larger projects. The efforts being made to broaden potential approaches to evaluation may have short-term benefits to provide workable assessment tools for the current environment, but will also provide insight into the effectiveness of each approach to be considered when regular in-person instruction begins again.

At Northeastern University in both a conservation principles course and a fluid mechanics course, students were assigned a short project in addition to each exam, with each project due date being within a week of its respective exam. The project allowed students the opportunity to present their conceptual understanding and knowledge in a creative format, being able to make a comic, video, or simulation addressing concepts from that part of the course. The projects provided a means for students to either reiterate their grasp on the course concepts in either a follow-up or lead-in to their performance on the exam, or to better show their understanding in a different, less-stressful, more open format in potential response to their previous performance. The introduction of the projects also served to reduce the number of exams in the semester, and could be completed remotely.

In this work, we discuss the impacts of these efforts in both short and long semesters (7 week vs. 14 week schedules), student performance and feedback, and insight into the advantages and disadvantages of this approach.

630363 Lessons Learned during the COVID-19 Pandemic: A Literature Synthesis

Abstract:

The COVID-19 pandemic brought a widespread shift in instructional practice as faculty scrambled to shift to remote instruction. One positive consequence is that this shift has inspired many instructors to think carefully about their teaching practice and the ways that technology tools and instructional strategies support their learning goals. Correspondingly, STEM educators have been investigating these needed changes and reporting them. There have also been coordinated efforts within our community to support chemical engineering educators such as five AIChE EdDiv Cirtual Communities of Practice. A corresponding spike in discipline-based education research (DBER) has also occurred. For example, over 250 papers have appeared in three archival journals related to chemical engineering: *Advances in Engineering Education*, *Biomedical Engineering Education*, and the *Journal of Chemical Education* (*Chemical Engineering Education* will be also publish a special issue in January 2022). Each of these reports focuses on specific contexts, mostly at the course level. In this paper we report a synthesis of the literature to identify how educators responded to the pandemic, and what we can learn from it. Topics analyzed include: course design, content delivery, assessment practices, classroom environment, project and laboratory work, the student experience, and social justice.

630756 Interventions to Mitigate Loss of Community during COVID-19

Abstract:

The switch to remote learning in response to the COVID-19 pandemic caught everyone by surprise, but faculty worked together to quickly transition. The administration worked to help student set up home learning stations by loaning them equipment and providing IT assistance. In terms of mechanics, everything seemed to be working. Students quickly learned, however, that webcams were overrated. Live-streaming video destabilized their already weak internet connections. Coupled with Wi-fi frustration is the requirement that you and your roommates must be dressed and the room be some semblance of clean. Students rarely turned on their cameras. For faculty this meant talking to a bunch of black squares with names. For students, they lost their peers.

In Spring of 2020, the loss of community was overlooked. Students' established study groups and communication channels for the semester were challenged by the campus closure. Though functional, academic life had become strange and unsatisfying. The Fall 2020 classes were a different experience. Although there were still technology challenges, support systems were in place to provide lectures, host discussions and even to perform lab experiments safely. The university was most concerned about the social interactions amongst the freshmen and worked vigorously to build relationships among peers. Seniors generally had well-established relationships with their classmates. Sophomores still had their freshman networks but were beginning to branch off into their selected majors. The juniors were most affected by the loss of community. Their friends who were in other majors did not understand the challenges they were facing with transport phenomenon or reactor design. Many juniors dropped their most challenging courses or withdrew from school to wait until in-person classes were available. An unusually large number of students landed on academic probation.

As associate chair for undergraduate studies, I reached out to all students on academic probation and invited them to set up weekly Zoom meetings with me. Two accepted my invitation in Fall 2020. During our 5- to 15-minute sessions they talked about their academic and home lives without judgment from me. I was able to point out resources they may have been unaware of, but generally tried to avoid giving unsolicited advice. I encouraged them to find something to look forward to each week. By the end of the semester, they had both completed all courses successfully and were no longer in academic jeopardy. I repeated this for Spring 2021 and had three students sign on with similar success. In the presentation I will look at common characteristics of those who chose to meet with me and why this was helpful to them.

In Spring of 2021, the lack of community was most noticeable. In one junior-level course, the instructor and course content had not changed, but the number of students struggling with the course had increased dramatically. The number of students complaining about the workload or lack of support in all of our junior-level courses had increased. In conversation with these struggling students it seemed that they were suffering from an absence of people to hold them accountable. Nobody was asking those simple questions like "Where were you last Tuesday?" or "Have you started on the homework yet?". It was apparent that the juniors had not formed study groups within their major before the pandemic hit. In cooperation with the Omega Chi Epsilon chapter, the department established FLASH sessions (Facilitating Learning Activities with Student Help). Senior members of the honor society open a classroom for an in-person study hall. Occasionally the students who attend are looking for help with homework. More often, they just crave conversation. Although attendance has been small, these sessions have made an impact on the students that attend.

In this presentation, I will discuss these two types of intervention and how we have continued these into Fall 2021.

624914 Oral Exams in Undergraduate Cheme Thermo Taught Online

Abstract:

COVID-19 pandemic challenged educators in many ways, and in particular with a need for new means of students' assessment. Instead of trying to modify conventional written tests with advanced proctoring technologies, we proposed an alternative exam format, and tested it within the undergraduate Chemical Engineering Thermodynamics II course, taught synchronously online. The assessment was based on two parts: a "take-home" written part, somewhat similar to conventional written tests, and an oral defense of the written part. The oral part of the exam was carried out as individual WebEx meetings with students, and included a number of concept questions on the exam problems. The weight of the oral part in the overall exam grade was comparable to that of the written part. The effectiveness of this approach was demonstrated based on the consistency of the students' grades with other means of assessment, and comparison with the student performance in previous years. Additionally, the approach was evaluated based on several student surveys, which showed that many students liked this format, and found oral sessions helpful to feel more involved with the course. Furthermore, the vast majority of the students found this format of the exam helpful to learn the course material. Although this approach does not save the instructor's time spent on assessment, the students benefit from this time more. We conclude that our approach is a viable assessment option for classes below 20 students, or larger classes if teaching assistants/proctors are available. This approach can be used in the courses taught both online and in-person.

624774 Online Cheating: A Darker Side of Educational Technology

Abstract:

While technology has had many positive impacts on education, particularly during the COVID-19 pandemic, it has also offered new opportunities for academic integrity violations. In particular, "tutoring" websites such as Chegg and Course Hero can be used by students to post homework or even exam problems for another person to solve; these problems and solutions are then posted online, possibly violating copyright laws as well as codes of student conduct at the students' academic institutions.

This study examines the prevalence of custom instructor-created exam problems posted to tutoring websites without instructor knowledge. Findings will include the frequency of exam problems appearing online and whether the nature of the course (traditional, online, etc.) has any effect on frequency. Best practices toward (a) identifying whether students have posted materials online and (b) minimizing opportunities for students to post course materials online will be presented.

622899 Aiche Education Division Virtual Communities of Practice (VCPs): Updates from the 2020-2021 Academic Year

Abstract:

Since the campus closures associated with the COVID-19 pandemic began in March 2020, the AIChE Education Division has provided a supportive network for faculty development through the initiation and implementation of the Virtual Communities of Practice (VCP) Program. This program consisted of five communities focused on different areas of the chemical engineering curricula: (1) Material & Energy Balances / Thermodynamics, (2) Transport / Separations, (3) Reactors / Controls / Computing, (4) Design, and (5) Lab. Detailed descriptions of these communities, including survey analysis and results obtained from March to December 2020, have been previously published in *Chemical Engineering Education*.

The VCP program continued throughout the Spring 2021 semester and an additional survey was administered to participants to further study the impact of this program on different aspects of faculty professional development. Results aggregated from over more than one year of VCP meetings will be provided to highlight the impact of this program on the AIChE community. Plans for future iterations of the VCP program and associated faculty community-building activities will also be discussed.

623386 Modeling the Progression of Fibrosis with Dysregulation of ACE2 in COVID19 Patients

Abstract:

The severity of the COVID19 pandemic has created an emerging need to investigate the long-term effect of infection in healing patients. Many individuals are at risk of suffering pulmonary fibrosis due to pathogenesis of lung injury and impairment of the healing mechanism. Fibroblasts are the central mediators of extracellular matrix deposition during tissue regeneration. In the later phase of infection, tissue damage causes the activation of latent anti-inflammatory cytokines (TGF-b), which recruits the fibroblasts [1]. The accumulation of fibroblasts in the damaged site and excess deposition of fibrillar collagen lead to fibrosis. Additionally, SARS-CoV-2 enters the host cells via binding its spike protein with the Angiotensin Converting Enzyme 2 (ACE2) receptor, which is a key component in modulating the balance of the renin-angiotensin system (RAS). The dysregulation of ACE2 by the viral infection can shift the balance of RAS towards pro-inflammation and pro-fibrosis.

We have developed an open-source, multi-scale tissue simulator that can be used to investigate the mechanisms of intracellular viral replication, infection of epithelial cells, host immune response, and tissue damage [2]. Our model simulates the dynamics of ACE2 and fibroblast-mediated collagen deposition to account for the fibrosis at the damaged site in response to immune response induced tissue injury. The severity of infection and collagen deposition depends on the anti-inflammatory cytokine secretion rate, multiplicity of infection, and contact time for a CD8+ T cell to kill an infected cell. Using the model, we predicted the dynamics of fibroblast and collagen deposition for 10 days of post-infection in virtual lung tissue. Our result shows that collagen area fraction increases from 31% to 69% with the severity of infection. We use the ACE2 dynamics as input in a separate model of RAS to predict the change in RAS signaling and collagen deposition from homeostasis for normotensive and hypertensive patients. Our model also reveals that the variation in available ACE2 due to age and gender can lead to significant change in inflammation, tissue damage, and fibrosis.

References

1. Zhang, H. et al. "Effects of transforming growth factor-beta 1 (TGF-β1) on in vitro mineralization of human osteoblasts on implant materials." *Biomaterials* 12 (2003): 2013-2020.
2. Wang, Y, et al. "Rapid community-driven development of a SARS-CoV-2 tissue simulator." *BioRxiv* (2020).

Acknowledgment

This work was supported by the National Institutes of Health grant R35GM133763 and the University at Buffalo.

631819 DNA-Directed Patterning to Validate a Liposome Model of Sars-Cov-2

Abstract:

INTRODUCTION

Lipid-based nanoparticles, or liposomes, have found wide applications in combatting the COVID-19 pandemic. Most notably, lipid-based nano-formulations encapsulate mRNA in the recently-approved COVID-19 vaccines from Moderna and Pfizer-BioNTech [1]. As enveloped viruses have been shown to fuse with liposomes [2], lipid-based nanoparticles have been suggested as models of enveloped viruses themselves [3, 4]. Based on their ability to display surface antigens, liposomes have been used to model the immune response to antigens [5] and to trigger a T-cell response to SARS-CoV [6]. These lipid-based nanoparticle applications arose from the demands of an urgent health crisis and necessitated high-throughput, highly flexible methods for nanoparticle characterization.

We have developed a novel, high-throughput method, DNA-directed patterning [7], to validate lipid-based nanoparticles. This method relies on lithographically patterning single-stranded oligonucleotides on a glass substrate and subsequently hybridizing the complementary oligonucleotides to which liposomes have been tagged, ultimately leading to the controlled and precise patterning of liposomes. Through the development of S-liposomes—an ideal, safe model of SARS-CoV-2 that preserves its most basic features and function as a lipid bilayer nanoparticle that displays the SARS-CoV-2 spike (S) protein—we demonstrate the rapid characterization that DNA-directed patterning provides [8].

EXPERIMENTAL

Liposomes were fabricated by rehydrating lipid thin films (dioleoyl phosphatidyl choline, dioleoyl phosphatidyl ethanolamine, dioleoyl phosphatidyl glycerol acid, and cholesterol in a 4:3:2:7; 1,2-dioleoyl-sn-glycero-3-((N-(5-amino-1-carboxypentyl)iminodiacetic acid)succinyl) (nickel salt) at 5 mol%.; DiO, DiI, or DiD at 1 mol%) followed by sequential extrusion and labeling with either a His-tagged SARS-CoV-2 spike protein or a His-tagged CD63.

Amine-terminated single-stranded 20 nt DNA oligos were immobilized on glass substrates by selectively exposing regions of a Shipley 1813 photoresist-coated slide to UV light using a photomask and dropcasting the appropriate oligonucleotide solution. Following reductive amination with 0.25% sodium borohydride, photoresist could be stripped and reapplied to iteratively pattern subsequent oligonucleotide sequences for multi-component patterns.

Liposomes were labeled with an anchor DNA oligonucleotide containing a cholesterol molecule that intercalated in the membrane as well as an adaptor sequence that facilitated hybridization with the surface patterned oligo. This anchor was stabilized using a complementary co-anchor oligo-cholesterol molecule.

Liposome-associated cholesterol-oligonucleotide tags were hybridized to complementary oligonucleotides patterned on a glass substrate surface. To verify the presence of spike protein on the liposomes, biotinylated ACE2 (ACROBiosystems) in 2% bovine serum albumin (BSA, Sigma-Aldrich) in PBS was introduced and incubated for 1 hr at room temperature followed by Cy5-labeled streptavidin (Invitrogen) in 2% BSA (incubated for 45 minutes at room temperature). For neutralization assays, prior to ACE2 incubation, different concentrations of anti-SARS-CoV-2 receptor binding domain neutralizing antibody (range: 0-100 μg mL^{-1}; Human IgG1, ACROBiosystems) in 2% BSA were incubated with the patterned liposomes for 1 hr. Internalization assays were performed by incubating liposomes with 293T cells engineered to express ACE2 followed by time-lapse microscopy.

RESULTS AND DISCUSSION

Following the fabrication of S-liposomes, we utilized DNA-directed patterning to validate our ability to fluorescently label and visualize these lipid nanoparticles. We mixed three labeled lipid-based nanoparticles and flowed the mixture across the pre-patterned substrate. The complementary oligonucleotide tags hybridized to the patterned oligonucleotides, leading to distinctive patterns of labeled liposomes.

To verify that the S trimer was oriented correctly using DNA-directed patterning, we patterned fluorescent S-liposomes onto a glass substrate in a 50 x 50 array of squares, each 141 μm^2 in size. To verify binding specificity, we measured fluorescence for conditions with intermediate components removed (i.e. no oligonucleotide tag, liposomes without the S protein, no addition of biotinylated ACE2). The condition which included all components showed significantly higher fluorescence ($p<0.0001$, one-way ANOVA with Tukey's multiple comparison test) than those of the controls in which each component of the binding chemistry was systematically removed.

Using S-liposomes and CD63-liposomes labeled with different oligo sequences and dyes, we showed that specificity of Watson-Crick base pairing allowed for quick, specific, and simultaneous separation of the tagged liposomes to distinct patterns. After patterning, the introduction of biotinylated ACE2 followed by Cy5-streptavidin allowed us to measure relative ACE2 binding, which occurred only in the presence of S-liposomes and not in the presence of CD63-liposomes ($p < 0.0001$).

We then employed DNA-directed patterning to assess the ability of a commercially available neutralizing antibody (clone: HTS0483) to interfere with ACE2 binding. We patterned fluorescent S-liposomes and incubated them with different concentrations (0-100 $\mu g\ mL^{-1}$) of neutralizing antibody. As expected, the concentration of neutralizing antibody inversely correlated with Cy5 intensity.

To confirm that S-liposomes, as validated by DNA-directed patterning, exhibit biological function, we investigated the ability of ACE2-expressing cells to internalize these lipid nanoparticles. We seeded cells transduced to express ACE2 (293T+ACE2) and visualized internalization of fluorescent liposomes without surface protein (plain) or with S or CD63. To compare the effects of the neutralizing antibody in this system against our DNA-directed patterning assay, we next added a range (0-10 $\mu g\ mL^{-1}$) of neutralizing antibody concentrations with fluorescent S-liposomes to 293T+ACE2 cells. As expected, we found that increasing concentrations of antibody led to decreasing fluorescence, indicating successful measurement of antibody interference with the interaction between the S-liposomes and ACE2.

CONCLUSION

In conclusion, we have demonstrated a flexible, easy-to-use, high-throughput method—DNA-directed patterning—to validate and characterize lipid-based nanoparticles, in this case, a model of SARS-CoV-2. The flexibility of DNA-directed patterning of these lipid bilayer nanoparticles enables high-throughput studies and/or screening capabilities against liposomes displaying different surface proteins or variants of a surface protein, all simultaneously on a single substrate. An attractive feature of our method is that it uses entirely off-the-shelf components—oligonucleotides of any 20-nucleotide sequence can be readily purchased. Moreover, multiple surface proteins that could be used to target liposomes containing therapeutics or the release of such therapeutics could be characterized or tested in a multiplexed way. Taken together, S-liposomes and DNA-directed patterning can readily be adapted and customized to facilitate the development of new diagnostics and therapeutics for SARS-CoV-2. Looking beyond the current pandemic, DNA-directed patterning of liposomes can be applied in the rapid formulation and implementation of a wide variety of custom assays for rapid implementation and read-out, including competitive binding, drug targeting, and particle-cell interactions.

REFERENCES

[1] H. Ledford, *Nature* **2020**, DOI 10.1038/d41586-020-03593-7.

[2] S. Nir, N. Düzgünes, M. C. Pedroso De Lima, D. Hoekstra, *Cell Biophysics* **1990**, *17*, 181.

[3] F. Martin, R. Macdonald, *Nature* **1974**, *252*, 161.

[4] W. C. K. Poon, A. T. Brown, S. O. L. Direito, D. J. M. Hodgson, L. L. Nagard, A. Lips, C. E. MacPhee, D. Marenduzzo, J. R. Royer, A. F. Silva, J. H. J. Thijssen, S. Titmuss, *Soft Matter* **2020**, *16*, 8310.

[5] W.-Y. Wholey, J. L. Mueller, C. Tan, J. F. Brooks, J. Zikherman, W. Cheng, *Bioconjugate Chemistry* **2020**, *31*, 685.

[6] S. Ohno, S. Kohyama, M. Taneichi, O. Moriya, H. Hayashi, H. Oda, M. Mori, A. Kobayashi, T. Akatsuka, T. Uchida, M. Matsui, *Vaccine* **2009**, *27*, 3912.

[7] O. J. Scheideler, C. Yang, M. Kozminsky, K. I. Mosher, R. Falcon-Banchs, E. C. Ciminelli, A. W. Bremer, S. A. Chern, D. V. Schaffer, L. L. Sohn, *Sci. Adv.* **2020**, *6*, eaay5696.

[8] M Kozminsky, TR Carey, LL Sohn. "DNA-directed patterning for versatile validation and characterization of a lipid-based nanoparticle model of SARS-CoV-2." ChemRxiv [Preprint] (2021). DOI: 10.26434/chemrxiv.14208455.v1

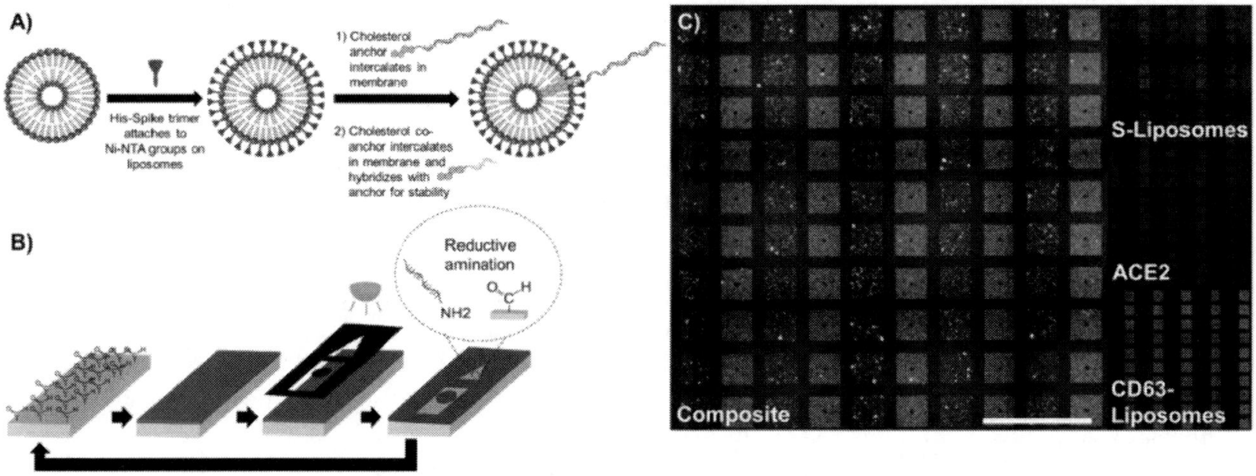

Figure 1: Fabrication and DNA-directed patterning of a liposome model of SARS-CoV-2. A) Liposomes are labeled with the SARS-CoV-2 spike protein and tagged with single-stranded (ss) oligonucleotides conjugated to a cholesterol molecule (orange) via the intercalation of that cholesterol molecule into the lipid bilayer. B) DNA-directed patterning of liposomes is first accomplished by patterning ss oligonucleotides using traditional photolithography. Photoresist is spin-coated and baked on aldehyde-coated substrates, patterned with a photomask using UV light, and then developed. Reductive amination of exposed aldehyde groups on the patterned substrates with amine-terminated ss oligonucleotides is subsequently performed. Photoresist is stripped with acetone for additional patterning of orthogonal ss oligonucleotides. C) S-liposomes (green) and CD63-liposomes (cyan) were patterned in alternating columns of an array of 141 μm by 141 μm squares, after which biotinylated ACE2 and fluorescent streptavidin were introduced. Scale bar = 500 μm. Adapted from Ref 8.

632826 Polymersomes Decorated with Sars-Cov-2 Spike Protein Receptor Binding Domain Elicit Robust Humoral and Cellular Immunity

Abstract:

Motivation: Since the emergence of the novel coronavirus SARS-CoV-2, researchers across the world have sought to rapidly develop vaccine candidates. The primary antigen targets are the Spike protein or its constituent receptor binding domain (RBD) which is responsible for ACE-2 binding and viral entry into the host cell. While several highly potent neutralizing antibodies against RBD have been isolated, RBD has also been shown to have lower immunogenicity than the full spike protein. Materials science and engineering approaches, particularly strategies involving nanotechnology, may improve RBD immunogenicity and are expected to aid in the development of new subunit vaccines.

Methods: We have previously reported the development of polymersomes (PS) that self-assemble from the oxidation-responsive block copolymer PEG-PPS and shown their efficacy in activating dendritic cells, inducing robust T cell immunity, and eliciting high antibody titers with broad epitope coverage. To further improve the humoral response of PS while retaining their ability to induce T cell immunity, we engineered their surface to display multiple copies of antigen, mimicking the physical form of the virus (Fig. 1a). We hypothesized that these surface-conjugated PS would result in the crosslinking of B cell receptors and subsequent production of neutralizing antibodies. Here, we evaluate RBD-surface-bound PS (RBD_{surf}) and RBD-encapsulated PS (RBD_{encap}) adjuvanted with MPLA PS compared to adjuvanted and unadjuvanted free RBD as vaccine candidates in mice.

Results: While all groups receiving adjuvanted RBD regardless of antigen formulation produced high titers of RBD-specific IgGs, only RBD-surface-conjugated PS elicited antibodies that neutralized live SARS-CoV-2 in vitro (Fig. 1b). These antibodies bound to linear epitopes found within the receptor binding motif, which is primarily responsible for interacting with ACE-2. Furthermore, RBD-surface-conjugated PS also showed robust antigen-specific germinal center responses and both $CD4^{+}$ and $CD8^{+}$ T cell immunity.

Implications: This work provides insights into differing B cell and T cell immunity of different antigen formulations in the context of COVID-19 and can thus be used to inform the design of next-generation subunit nanovaccines.

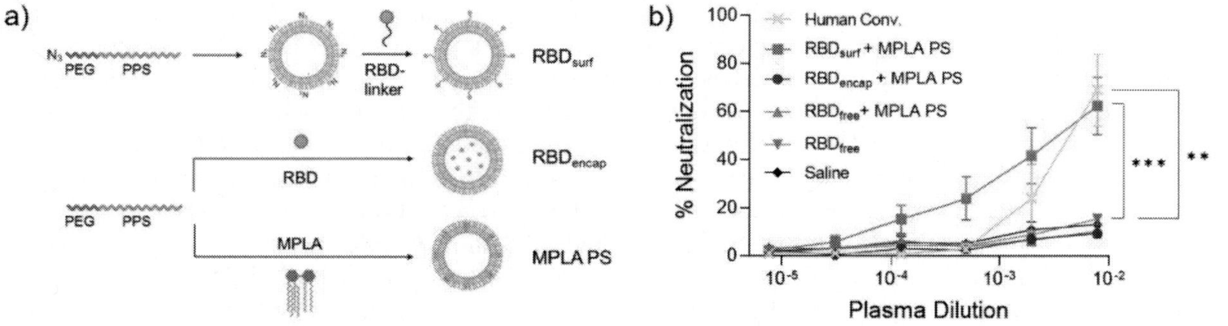

Figure 1. a) Schematic of polymersome formulation. b) Neutralization of live SARS-CoV-2 by plasma from vaccinated mice and human convalescent plasma.

622995 Inhalation Delivery of RNA-Loaded Lipid Nanoparticles Against Sars-Cov-2

Abstract:

Problem: Despite ongoing vaccination efforts, SARS-CoV-2 continues to present a significant global health burden. Furthermore, long-term immunity following vaccination is not guaranteed and the emergence of virus variants could further hinder the efficacy of this approach. These challenges suggest a continuing need for antiviral therapies that can reduce disease severity and lethality; however, available antiviral medications have shown limited impact on symptomatic disease. Efforts to identify drugs that are effective against SARS-CoV-2 have been complicated by its use of endogenous cellular components for viral replication. RNA interference (RNAi) is a promising alternative approach for antiviral therapy, where oligonucleotides with sequence homology to the viral RNA can selectively result in degradation of viral RNA transcripts and translational inhibition. RNAi sequences can be computationally determined based on well understood design rules; therefore, this approach could be rapidly adapted to different viruses or virus variants. Lipid nanoparticles (LNPs), a leading technology for RNAi delivery, protect RNA cargo from nucleolytic degradation and promote cellular uptake and endosome escape of RNA, leading to efficient gene knockdown. LNP formulations are usually administered through injection, whereas SARS-CoV-2 primarily infects epithelial cells in the respiratory tract, which are better accessed by topical administration via nebulization. A major obstacle in this delivery modality is the shear forces produced during nebulization can cause nanoparticle disruption and agglomeration, which may hinder the efficacy of nebulized LNPs. Here, we develop an RNAi strategy targeting SARS-CoV-2 and demonstrate conditions that stabilize LNPs during nebulization.

Methods: Lipid nanoparticles (LNPs) encapsulating RNA were prepared by microfluidic mixing, which enables production of particles that are about 100 nm in diameter with high RNA encapsulation efficiency (>90%). For nebulization studies, model RNA cargo was loaded in LNPs, which was either firefly luciferase-targeting siRNA (siLuc) or cyanine 5-labelled universal negative control siRNA (Cy5-NC). LNPs were diluted in buffer solutions of varying pH with or without selected excipients prior to nebulization. Nebulization of LNPs was achieved using an Omron MicroAir vibrating mesh nebulizer. Nebulizer output was collected in a conical tube and LNPs in this solution were evaluated in terms of hydrodynamic diameter, encapsulation efficiency and RNA recovery. Additionally, nebulized LNPs were assessed for cellular uptake, measured by flow cytometry (Cy5-NC-LNPs), and luciferase knockdown, determined by luminescence (siLuc-LNPs).

Computational tools were used to design a set of siRNAs that target SARS-CoV-2 transcripts. siRNA targets were inspected in the folded RNA transcript structure using RNAstructure software. To evaluate the potency of these siRNAs, plasmids were prepared that express enhanced green fluorescent protein (EGFP) flanked by the 3' and 5' untranslated regions (UTR) of SARS-CoV-2 under the control of a constitutive promoter. Plasmids representing the positive sense viral transcripts (5'UTR-EGFP-3'UTR) were chemically transfected into Vero cells, which are commonly used to grow SARS-CoV-2. Following drug selection, stably expressing uniform clones were identified by fluorescence microscopy and flow cytometry. Knockdown of viral RNA constructs could be quantified as a reduction in EGFP mean fluorescence intensity (MFI) measured by flow cytometry. siRNA against EGFP (siEGFP) dosed using lipofectamine or LNPs was used as a positive control in knockdown experiments to determine concentration (tested range 20 nM - 200 nM), duration and efficiency. Candidate siRNAs targeting the 5'UTR of SARS-CoV-2 were synthesized by Integrated DNA Technologies. The full set of designed siRNAs were evaluated by transfecting 50 nM with RNAiMAX Lipofectamine into transgenic Vero cells individually and in combination. Varying dosage concentrations (tested range 1 pM - 50 nM) were also evaluated. Promising siRNAs were encapsulated in LNPs prior to dosing Vero cells containing viral RNA-EGFP constructs. siRNAs were also evaluated in Vero cells infected with SARS-CoV-2 virus. Vero cells were treated with siRNA for 6 hours prior to viral infection. Sub-genomic and total viral RNA were quantified by quantitative polymerase chain reaction (qPCR) at 24 hours post-infection.

Results: Nebulization of siLuc-LNPs in phosphate buffered saline (PBS, pH 7.4) resulted in significant loss of siRNA encapsulation and more than a doubling in particle size. This confirmed that a stabilization strategy is needed for LNP nebulization. LNPs contain an ionizable lipid that is protonated at low pH. It was hypothesized that reducing nebulization buffer pH could impart particle stability by increasing surface charge and hence electrostatic repulsion between particles. Additionally, ionization of LNP lipids at low pH could increase retention of encapsulated RNA by maintaining electrostatic complexes between the ionizable lipid and the RNA phosphate backbone. Encapsulation efficiency was recovered to near pre-nebulization levels for LNPs nebulized at reduced pH. However, reduced nebulization buffer pH did not prevent LNP particle growth. Further, low pH buffer resulted in a reduction in total RNA recovery from the nebulizer, presumably due to adsorption of charged LNPs to the nebulizer plastic surface. To overcome these remaining issues, we evaluated several excipients to be included in the low pH nebulization buffer, including surfactants, sugars, and hydrophilic polymers. Of the screened excipients, we identified an excipient that largely prevented particle growth and significantly improved RNA recovery. We further validated that LNPs nebulized under these optimized conditions retained cellular uptake and efficient gene knockdown. Fluorescently labelled Cy5-NC-LNPs were taken up in Vero cells to a similar degree before and after nebulization. Additionally, siLuc-LNPs exhibited no significant change in luciferase knockdown potency when nebulized in the optimized conditions.

We computationally designed a set of ten siRNA oligonucleotides that target the UTR of SARS-CoV-2 RNA transcripts. Transgenic Vero cells expressing viral RNA-EGFP constructs were generated to evaluate siRNA potency. Generation of cell lines stably expressing plasmids was confirmed by observation of EGFP fluorescence by confocal microscopy and quantification of EGFP+ cells by flow cytometry. The cell line exhibiting the greatest proportion of EGFP+ cells (97%) was selected for evaluation of siRNAs. First, this system was validated by demonstrating efficient knockdown of the viral RNA-EGFP construct with siEGFP as a positive control. EGFP fluorescence was reduced by 68.2% or 56.2% when siEGFP was dosed with lipofectamine or LNPs, respectively. Next, we evaluated our set of ten candidate siRNAs in this viral construct model with transfection via lipofectamine. Of these siRNAs, five resulted in significant (>55%) EGFP knockdown. To test whether the siRNAs behave synergistically to knock down expression, we also transfected with combinations of the top candidates; this approach yielded no beneficial effect. Further studies evaluating different siRNA doses allowed the identification of three promising siRNA sequences for further study. These three siRNAs also exhibited efficient knockdown of viral RNA-EGFP constructs when encapsulated in LNPs. LNP delivery of as little as 10 nM siRNA yielded >60% knockdown. Importantly, in Vero cells infected with SARS-CoV-2 virus, pre-infection treatment with one siRNA candidate via lipofectamine resulted in a 2-log reduction in sub-genomic viral RNA compared to untreated cells.

Implications: Overall, inhalation RNAi therapy is a promising strategy to manage viral infections. We demonstrate an approach to stabilize LNPs during nebulization which will enable RNAi therapies against SARS-CoV-2 and other respiratory viruses. Insights gained from these studies may be applicable for stabilization of nanoparticle systems for pulmonary delivery. siRNAs designed in the present work showed promise through knockdown of model viral RNA constructs *in vitro*. Further studies will evaluate this delivery strategy for efficacy in an *in vivo* model of viral infection. The ability to effectively deliver potent RNAi via inhalation could be broadly clinically useful for a range of respiratory diseases.

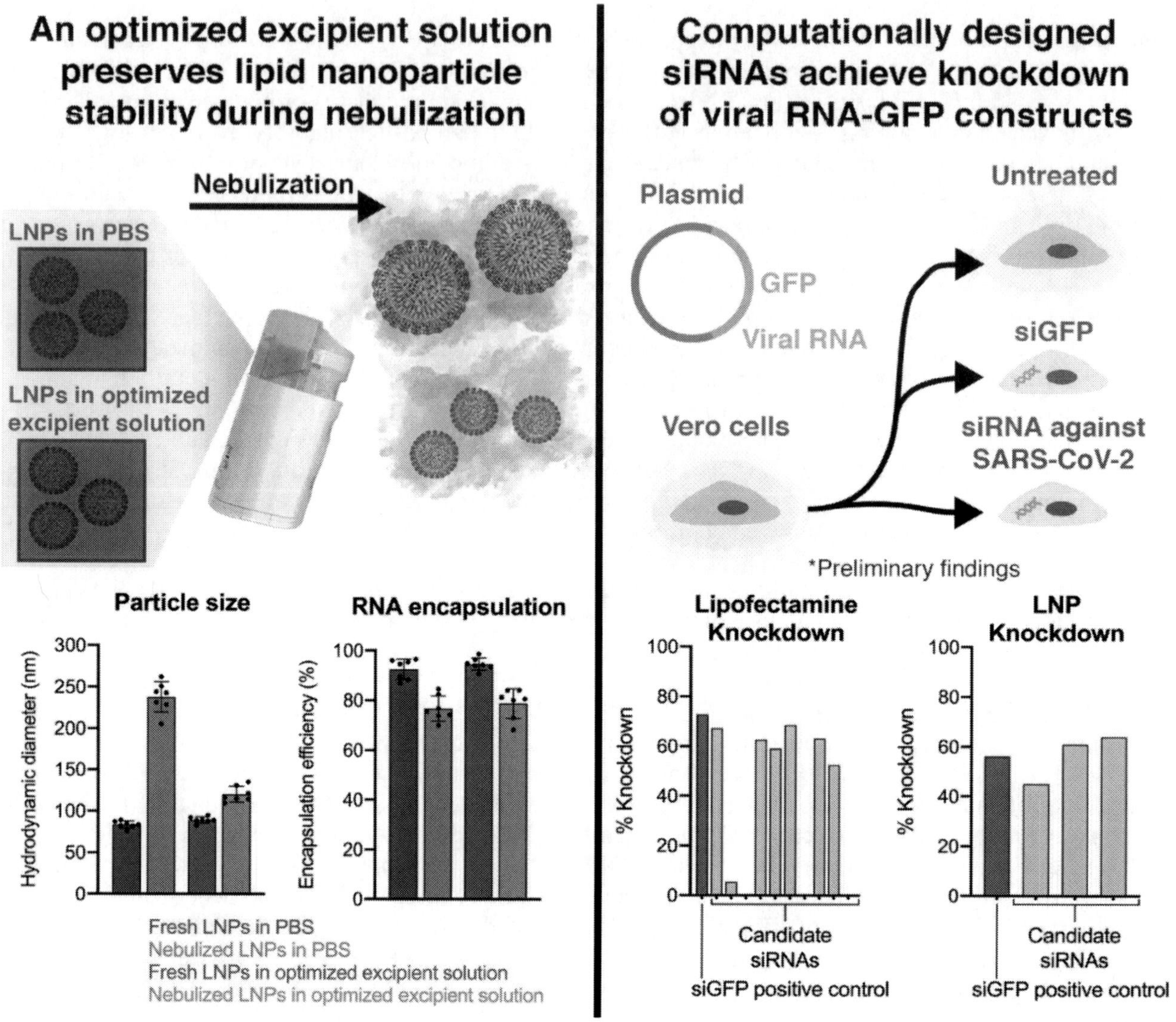
An optimized excipient solution preserves lipid nanoparticle stability during nebulization
Nebulization
LNPs in PBS
LNPs in optimized excipient solution
Particle size
Hydrodynamic diameter (nm)
RNA encapsulation
Encapsulation efficiency (%)
Fresh LNPs in PBS
Nebulized LNPs in PBS
Fresh LNPs in optimized excipient solution
Nebulized LNPs in optimized excipient solution
Computationally designed siRNAs achieve knockdown of viral RNA-GFP constructs
Plasmid
GFP
Viral RNA
Vero cells
Untreated
siGFP
siRNA against SARS-CoV-2
*Preliminary findings
Lipofectamine Knockdown
LNP Knockdown
% Knockdown
Candidate siRNAs
siGFP positive control

633881 A Human Pluripotent Stem Cell Approach to COVID-19 (Invited Speaker)

Abstract:

There is an urgent need to create novel models using human disease-relevant cells to study SARS-CoV-2 biology and to facilitate development of therapy. As *SARS-CoV-2 primarily infects the respiratory tract, we developed a lung* organoid model using human pluripotent stem cells (hPSC-LOs). The hPSC-LOs, particularly alveolar type II-like cells, are permissive to SARS-CoV-2 infection, and showed robust induction of chemokines upon SARS-CoV-2 infection, similar to what is seen in COVID-19 patients. Using hPSC-LOs, we performed a high throughput screen of FDA-approved drugs and identified entry inhibitors of SARS-CoV-2. Besides human lung models of COVID-19, we used directed differentiation of hPSC to establish a lung and macrophage co-culture system and model the host-pathogen interaction and immune response caused by SARS-CoV-2 infection. Our studies reveal differential immune response patterns in distinct macrophage phenotypes, which could be contributing factors to the spectrum of COVID-19 severity. Together, these data demonstrate that hPSC-cells infected by SARS-CoV-2 can serve as disease models to study SARS-CoV-2 infection and provide a valuable resource for developing COVID-19 therapeutics.

633882 Application of Biophysical and Chemical Engineering Principles for Understanding Molecular Scale Interactions Critical to Virus Entry and Infection of Its Host (Invited Speaker)

Abstract:

The coronavirus disease 2019 (COVID-19) pandemic has focused attention on the need to develop effective therapies against the causative agent, SARS-CoV-2, and also against other pathogenic coronaviruses (CoV) that have emerged in the past or might appear in future. Focusing on steps in the CoV replication cycle, in particular the entry steps involving membrane fusion that are vulnerable to inhibition by broad-spectrum or specific antiviral agents, is an astute choice because of the conserved nature of the fusion machinery and mechanism across the CoV family. For coronavirus, entry into a host cell is mediated by a single glycoprotein protruding from its membrane envelope, called spike (S). Within S, the region that directly interacts with the membrane is called the fusion peptide, FP. It is the physico-chemical interactions of the FP with the host membrane that anchors it, thus enabling the necessary deformations of the membrane that lead to delivery of the viral genome into the cell when a fusion pore opens. As chemical engineers, we contribute to this fundamental work by leveraging our understanding of thermodynamics, kinetics, and intermolecular interactions to describe FP interactions with the host membrane at the most fundamental molecular level to facilitate the development of strategies to limit those interactions to stop the spread of infection. In this talk, I will describe our work on understanding the impact of calcium ions on CoV infection. Using cell infectivity, biophysical assays, and spectroscopic methods, we found that calcium ions serve to stabilize the fusion peptide structure during conformational change that then allows its insertion into the host membrane, resulting in increased lipid ordering in the membrane. This lipid ordering precedes membrane fusion and has been shown to correlate with increased fusion activity, as higher extents of fusion are observed as calcium concentration increases, aligned with higher levels of infection in the presence of calcium. Finally, depletion of calcium ions leads to structure and activity changes that correlate well with *in vitro* experiments using calcium-chelating drugs. Under these conditions, cell infection dropped, pointing to the possibility of such drugs as therapeutic interventions.

635087 Commercialization of Aptamer-Based Electrochemical Sensors

Abstract:

QSM Diagnostics is in the process of commercializing electrochemical sensor technology developed by the Goluch Group at Northeastern University. The initial application of the technology is for rapidly diagnosing and monitoring bacterial infections in companion animals. This is accomplished by detecting quorum sensing molecules (QSMs) that are present in specimens collected from the animals. The platform technology, however, can be adapted to almost any target molecule of interest by simply exchanging the recognition element that is attached to the sensor. The QSM sensing platform uses aptamers as recognition elements in conjunction with square wave voltammetry. Theoretically, any properly developed aptamer can be integrated to create a custom rapid, quantitative, testing system.

As an example of this turn-key solution capability, we have now demonstrated the detection of moxifloxacin, a broad-spectrum fluoroquinolone antibiotic, in human serum samples. This presentation will describe the process by which the new sensor was developed and tested as well as the challenges associated with the development process.

635137 (Invited Talk) Developing Nanosensor Technology to Detect Cancer in Patients

Abstract:

We have been developing and translating a technology to detect biomarkers quantitatively in humans to enable the early diagnosis of ovarian cancer. There is no effective screening test for ovarian cancer and no reliable method of early detection. Approximately 85% of patients with ovarian cancer will have widely disseminated peritoneal disease at presentation, resulting in five-year survival rates of < 30%. This is in stark contrast to patients diagnosed with early-stage localized disease, where survival rates are over 90%. The early detection of ovarian cancer would markedly increase survival rates. Serum biomarkers have repeatedly failed to identify early disease or change mortality rates, in part due to low specificity. We believe that the measurement of biomarkers using an implantable device in the uterine cavity will allow for discrimination of benign states from malignant disease. We are working in collaboration with clinicians to develop an implantable sensor platform for protein biomarkers capable of minimally-invasive, highly-sensitive, reproducible, in vivo measurements in disease models. We demonstrated a device to implant the sensors in animals and resected human uteri. Our goal is to develop the capability for real-time and progressive patient-centered monitoring of biomarkers for early ovarian cancer detection in at-risk patients.

630605 (Invited Talk) Wearable Health Sensors

Abstract:

The need for well-managed hydration and the monitoring of physiological parameters in real-time is not only crucial for athletes to optimize performance but to also prevent illness and injury from occurring. Understanding sweat loss and calculating hydration requirements to avoid dehydration or overhydration is a significant challenge for athletes. This is because sweat losses vary significantly from athlete to athlete both in terms of concentration of electrolytes and total volume of sweat lost. When sweating, an individual may lose as little as ~200mg of sodium per liter or as much as ~2000 mg. This variance, combined with significant differences in sweat volume, lead to vastly different hydration requirements among individuals. The RooSense Optimal Sweat Performance (OSP) Band is the first lightweight fabric sensor to provide real-time information regarding hydration levels during exercise or training through selective determination of sodium levels, a better marker for athletic performance, safety, and injury.

Monitoring sweat is a compelling choice to gain insight into a person's health at the molecular level, as sweat is rich in physiological and metabolic information that can be obtained non-invasively. In spite of advances made towards the detection of biomarkers in sweat, there is no sensor capable of long-term detection in constricted or load-bearing applications where other flexible plastic sensors might cause discomfort. Textile-based and nanocomposite technologies, such as those developed at RooSense LLC, provide an optimum scaffold for monitoring biomarkers at the surface of the skin as they do not influence the sweat composition or sweat rate and can be seamlessly integrated for wearable sensing and translational medicine applications.

This presentation will focus on the development of a prototype wearable hydration monitor, validation of the prototype in the laboratory, on body testing in an exercise physiology laboratory.

630200 Resilience in the Face of COVID-19: Managing a User Program at the Molecular Foundry

Abstract:

Introduction

The Molecular Foundry is a Nanoscale Science Research Center, located at Lawrence Berkeley National Laboratory (Berkeley Lab) in Berkeley, California, sponsored by the Department of Energy (DOE). The Molecular Foundry is a knowledge-based User Facility, where leading-edge resources in instrumentation, analytical capabilities, and scientific expertise in nanoscale science are made available via a peer-reviewed proposal process. Roughly 1000 researchers or "users" come to the Foundry from around the world to conduct their work each year. The COVID-19 pandemic presented challenges and opportunities in conducting collaborative on site research. This presentation focuses on the operation of the user program amid the pandemic, with an emphasis on strategies for maintaining research productivity and user access. We also present an outlook on ways that gains in virtual collaboration and science during the pandemic can provide opportunities in a post-pandemic world.

The Molecular Foundry provides scientific expertise and capabilities distributed among seven facilities: Inorganic, Organic, and Biological facilities for synthesis, preparation, and assembly; Nanofabrication, for processing and integration; the National Center for Electron Microscopy and Imaging and Manipulation, for characterization; and Theory, for understanding and predicting material properties.

The Molecular Foundry maintains a highly collaborative research model. Staff scientists and technical staff at the Molecular Foundry interact with users through instrument training, consultation during experiments, support with data interpretation, and collaborative writing of journal articles. For each 12 month project, users stay in the local area for an average of three months while they work on site.

Spring 2020 Shelter-in-Place

In spring 2020, the scale and urgency of the COVID-19 pandemic in the United States became apparent, and the city of Berkeley enacted a Shelter-in-Place order effective March 17, 2020. The Molecular Foundry responded by pausing research operations on site while keeping research facilities safe and ready for reopening. This directly impacted every one of the hundreds of research projects that were active at the time, and presented a challenge to the Molecular Foundry's commitment to safety and user access.

The Foundry staff collaborated to respond to this challenge with a holistic approach to keep science moving forward safely. All user projects received a 12-week extension, to account for this pause on work. During this period, the Foundry developed protocols for safe work on site, aligned with requirements for Berkeley Lab and applicable regional, state, and national guidelines. These included the use of face coverings, hand-washing, distancing/density control, and surface disinfection.

While maintaining a very low density of people on site, a few high-priority projects were allowed to proceed, including research on COVID-19 and SARS-CoV-2 detection, treatment, and transmission mitigation. These early research activities served as a testing ground and model for the further expansion of access in the months to follow. Staff from across all seven facilities began to explore investment in virtual modes for training, user support, and instrument access amid the pandemic. Berkeley Lab developed new training modules for safe work on site during the COVID-19 pandemic, launched a lab density study to determine appropriate safe occupancy limits for all laboratories across the site, and developed a Health Pledge,symptom check system, and a specific COVID19 training module that was required for all on site personnel.

Summer 2020 and Beyond

The months that followed the onset of the pandemic were characterized by continual adaptation to the emerging pandemic. As of the writing of this abstract, this phase continues to evolve as vaccination campaigns are gaining momentum nationally. These are some of the strategies employed to manage the user program during this phase of the pandemic:

Communication

Communication with the user community and the staff of the Molecular Foundry was a central pillar of maintaining safe work at the Molecular Foundry. This included email messages to all users generally, frequent discussions among staff across all roles, and communication with Berkeley Laby leadership.

Shift Work to Manage Density

Upon emerging from the full shelter-in-place order, moderate numbers of researchers were allowed to work on site at the Molecular Foundry. The total number of persons accessing Berkeley Lab was managed centrally by the Laboratory, and the Molecular Foundry worked with an allotment of "slots" for site access throughout the duration of the pandemic restrictions. A three-shift model was deployed. Slots were distributed among the seven facilities, and researchers requested a slot ahead of time, subject to approval by Molecular Foundry leadership and compliance with the Health Pledge, weekly symptom check, and a specific COVID19 training requirements.

Prioritizing Research Projects

Facing limited access compared to normal operations, it was necessary to consider carefully which staff and user work could be accomplished on site. The Molecular Foundry values equity of access for users, held in balance with safety precautions. We aimed to carefully consider such factors as the potential to impact the fight against COVID-19, DOE and Berkeley Lab research priorities, inherently safe activities, experimental complexity, travel considerations, and user training.

Badged Entry for Approved Work

During the pandemic, the Molecular Foundry implemented badge-reader entry for all buildings. Coupled with weekly calendars and daily posting of all persons with prior approval to enter the buildings, this provided a layer of accountability around access to the spaces.

Regular Safety Meetings

Berkeley Lab implemented weekly virtual meetings focused on COVID-19 safe operations and other safety and space planning topics. These meetings require attendance by at least one staff representative from each of the seven facilities, and are open to all Molecular Foundry staff and users. Each facility hosted a daily safety check-in for work planning.

Carefully Considered Training

Training presented a particular challenge during the pandemic, as many users receive training on complex tasks at the Molecular Foundry, traditionally accomplished side-by-side. After several months, the Foundry formalized guidelines to consider for new user training, considering such factors as lab occupancy, the potential to develop tools to support distanced training, inherent hazards in the proposed work, and robustness of the instruments.

Development of New Remote and Virtual Training and Access

Molecular Foundry staff rallied around the call to support user access virtually. Specific responses are adapted to the needs and opportunities of each facility. A sampling of tools include: training videos, holographic augmented reality for training, audio headsets to support distancing, remote operation of select equipment, remote login for data processing, virtual presence via video conferencing during experiments, telepresence robots, and a variety of informal communication and collaboration tools.

Calls for Proposals Scope

Two of the standard calls for proposals have been fully within the pandemic. While we kept the calls open for the most exciting research across nanoscience, we balanced the practical considerations with some emphasis for these calls. We encouraged users to submit proposals that included COVID-19 research, theoretical and computational work, remote access to instruments, or data collection for remote analysis, and proposals requesting samples of materials developed at the Foundry.

Expanded Definition of Remote Users

The Molecular Foundry is one of five Nanoscience Research Centers sponsored by the DOE, and therefore shares certain reporting parameters for our user programs. Effective for FY 2020 reporting, the centers collectively adopted an expanded definition of a remote user, to recognize the increasing role of remote access for science at these knowledge-based user facilities.

Virtual Meetings and Conferences

Meetings have shifted online, supported by zoom video conferencing. This includes smaller team meetings, larger seminars spanning the Molecular Foundry, and our annual User Meeting--a nanoscale science research conference which drew 900 registrants in 2020.

Outcomes and Future Outlook

The pandemic presented challenges to maintaining a strong nanoscience research user program. Nevertheless, overall metrics reflect resilience in the face of these challenges. The User Program remains strong, serving 643 on-site and 97 remote users in FY 2020, according to formal reporting definitions, with a total of over 1500 named researchers on active projects during FY 2020. Demand for the user program remains strong, with around 225 standard proposals received during the most recent spring 2021 call for proposals, roughly two thirds our pre-COVID maximum. Research productivity as measured by journal articles has remained robust when comparing 2020 to 2019 and 2018.

The shift to increased virtual access presents opportunities to expand the geography of our user base to include more researchers outside our local area, due to reduced barriers in travel once post-pandemic travel resumes. We have seen an increase in sample-only proposals, enabling new ways of working together across institutions. We also have potential to include more people in the active experimental data collection phase, as undergraduate researchers and principal investigators increasingly log in remotely to participate in experiments that previously were completed by just one member of the research team. Furthermore, shifting our annual User Meeting online has facilitated a tripling of conference registrations, with potential to reach new communities with our science.

As we look ahead to 2022 with optimism, we hope to learn from the unique challenges presented by the pandemic. We will have new facets of this experience to manage, including a safe return to work for our users. The Molecular Foundry remains committed to advancing user science to bring solutions to the world.

629703 Global Supply Chain Optimization for COVID-19 Vaccines Under Covax Facility

Abstract:

With the pandemic in action, over 130 million people across 219 countries have been infected by COVID-19. The number of mortalities due to COVID-19 has climbed to around 3 million [1]. Though around 80% of the infected patients have recovered, several patients are becoming reinfected recently appeared. As a result, efforts for COVID-19 vaccines have accelerated around the world. The world's top pharmaceutical companies and research organizations have developed multiple vaccines. It is expected that the transmission can be slowed considerably if a sufficient population is vaccinated. Though a typical vaccine takes around ten years for approval, several COVID-19 vaccines have already been developed and approved in record times. As of 31 March 2021, around 750 million people worldwide have been vaccinated [2]. With world governments aggressively supporting vaccine developments and procurements, the vaccination drive is progressing rapidly.

However, to end the pandemic, we not only need COVID-19 vaccines, but the development of vaccines must be followed by a fair, fast, and efficient global distribution. The current challenge is the timely delivery of the vaccines. The expected volume of the vaccines per month adds up to around 850 tons and is more than the UNICEF suppliers' capacity combined [3]. Two main concerns regarding the global COVID-19 vaccine distribution are 1) Limited supply of resources for the delivery of the vaccines, e.g., cold chain equipment, and 2) Vaccine nationalism, i.e., governments signing agreements with companies to procure surplus vaccines. To overcome these challenges, the COVAX facility was launched by WHO for multiple vaccines to prevent vaccine nationalism. It aims to ensure fair access to COVID-19 vaccines for all countries, irrespective of their financial status. It allows countries to pool resources for effective vaccine development and distribution. Under COVAX, deals for vaccine procurement have already been made. However, vaccine distribution is still under development. It faces various challenges due to (1) the variety of vaccines, (2) person-specific doses, 2) unknown country-specific demands required for herd immunity threshold, especially low-middle income countries, (5) low-temperature storage protocols for most vaccines, and (6) unanticipated disruptions in both production and distribution.

Addressing these challenges requires a multi-prong approach that must go beyond straightforward pharmacological measures. The two primary goals are to (1) estimate the herd immunity threshold and (2) effect an efficient global vaccine distribution. In this work, we explore the time-varying nature of reproduction number during the first wave of the

COVID-19 pandemic using confirmed cases, recovered cases, fatalities, and timings of lockdown phases. We propose a mean-field epidemiological model for COVID-19 that extends the classical SEIR model [4]. We describe NEIASCFR, an 8-compartmental model with variable transmissibility over time, by considering the time-varying nature of parameter values over different lockdown phases. We estimate our model's parameters with public data on confirmed cases, recovered cases, and fatalities for each country. The parameter values are used to determine reproduction number and herd immunity threshold. The model is applied to study the herd immunity threshold of 9 countries (India, Nigeria, Iran, Pakistan, Thailand, Mexico, Democratic Republic of Congo, Egypt, Thailand), constituting 50% of demand under the COVAX facility.

Using this within the framework of COVAX, we develop a global supply chain model to optimize the production and allocation of vaccines to participating countries from various production centers. The supply chain was optimized considering potential disruptions, vaccine approval dynamics, and vaccine candidates' approval with different risk mitigation strategies. Given: 1) Network configuration consisting of production facilities and different regions. 2) Initial vaccine procurement quantities. 3) Capacity of different storage containers. 4) Costs of procurement, storage, transportation, and risk mitigation per unit. 5) Different scenarios, we determine 1) Production quantities of vaccines at different centers and vaccine allocation to each demand region. 2) Extra procurement of vaccines at each center. 3) Cost of procurement and transportation.

References:

1. Worldometer. COVID-19 CORONAVIRUS PANDEMIC. 2021; Available from: https://www.worldometers.info/coronavirus/.

2. Bloomberg. Covid-19 Tracker. 2021; Available from: https://www.bloomberg.com/graphics/covid-vaccine-tracker-global-distribution/.

3. UNICEF, UNICEF outlining plans to transport up to 850 tonnes of COVID-19 vaccines per month on behalf of COVAX. 2020.

4. Li, M.Y. and J.S. Muldowney, Global stability for the SEIR model in epidemiology. Mathematical biosciences, 1995. 125(2): p. 155-164.

632079 Sustainability and Resiliency Assessment for Industrial Synergies between Renewable Energy and Chemical Production

Abstract:

Photovoltaic (PV) is the fastest growing energy technology and is expected to maintain this growth in the coming years. Due to the durability of PV modules, there is a lengthy time gap between installation and decommissioning of the modules, and it is expected to create large scale waste worldwide in the near future. On the other hand, chemical manufacturing systems are undergoing significant transition towards bio-based production to meet the goal of sustainable low carbon feedstock. This work presents a potential synergy between renewable energy wastes and bio-based polymer production through extending a previously proposed system to circularize the PV lifecycle by diverting organic byproducts from PV recycling to agricultural production. We identify and introduce additional synergies as links in the network to close the loop between PV module disposal and manufacturing. The chemical outputs from the PV recycling process are used as fertilizers for crops that are potential feedstock for biopolymer production. These biopolymers are used in the manufacturing of PV backsheets, replacing the current petroleum-based products. We quantify the total life cycle environmental and economic benefits at scale of implementing this synergy along with total reduction in GHG emissions. To evaluate the design, we also use a network-based approach to understand the structure and topology of the proposed system by exploring various network measures as well as information theory-based measures, to assess resilience and long-term sustainability. We show quantitatively that network measures like diameter and density are ineffective at scoring the network resilience due to the nature of these measures in original network science literature, while centrality measures are useful for finding the relative importance of certain nodes, which may highlight network vulnerabilities that factor into the overall resilience. Likewise, ascendency and reserve values for the network appear useful in determining the overall resilience and sustainability of the network, however, based on their application to comparing two alternative network scenarios, these metric interpretations should be more rigorously defined for industrial networks before being incorporated for industrial network design.

633441 A Community-Level Energy Planning Model for the Dynamic Optimization of Waste-to-Gas and Waste-to-Power Anaerobic Digestion Supply Chains.

Abstract:

Society has growing concerns over the long-term environmental impacts of climate change and the accumulation of organic waste in local communities. Community energy planning is increasingly focused on the use of novel technologies to provide sustainable and cost-effective heat and power. Furthermore, changes in energy demand patterns support the development of dynamic optimization models of energy dispatch. This study focuses on the use of anaerobic digestion (AD) technology for converting human, agricultural, and industrial waste into renewable natural gas (RNG) and power. We develop a dynamic mixed integer linear programming (MILP) in Calliope to optimize the production and dispatch of RNG and power.
Anaerobic Digestion (AD) is a microbiological process that is used in processing organic, biodegradable wastes and side streams into methane-rich biogas and nutrient-rich digestate. Previous techno-economic (TEA) and lifecycle (LCA) analyses indicate that AD can be a commercially competitive and sustainable technology for communities to generate clean energy. In this study, we evaluate the use of AD in Ames, Iowa to convert human and animal waste, biomass, and wastewater into RNG and power. RNG and power are consumed locally to provide industrial, commercial, and residential heat and power demand. The AD process generates by-product digestate that is returned to Iowa farms. The model covers simulates 1 year of energy demand in hourly and monthly increments.

The MILP identified 13 locations in the vicinity of Ames, Iowa capable of producing power or RNG. The electricity price ranged between \$0.065 - \$0.18/kWh and the RNG price is estimated at \$0.026/kWh (\$7.62/mmbtu). Electricity production varied between 30 and 50 million kWh per month. Gas production varied between 15 and 40 million kWh per month. All the electricity demand in Ames, Iowa could be supplied from local waste resources. Summer gas demand can be supplied locally. However, the gas supply will need to be supplemented with gas from the external grid to meet winter demand. Technological improvements and public incentives could increase the adoption of AD in community energy planning systems. Further research will evaluate the impacts of variabilities in energy supply and demand to address grid disruptions.

625575 Asset Integrity Management (AIM) for Hydrogen: Major Incident Risk Reduction By Integrating a Risk-Based Safety Approach

Abstract:

Asset Integrity Management (AIM) is a well-understood process for optimizing the management of assets over their whole life-cycle. If applied correctly, AIM can offer asset owners and operators the ability to manage risk and enable the assertion, with confidence on an evidential basis, that their assets are safe and reliable. For AIM to be most effective, the assets' design must consider the whole life-cycle of the assets' operation. Adopting a risk based approach to safety and operations as early as possible in the design phase of a facility is demonstrated to improve long-term reliability, availability and maintainability. Combining a risk based approach at the design stage with AIM during operations has the potential to simplify and align processes. Additional benefits include the improvement of long-term reliability, availability and maintainability. By integrating a risk based approach with AIM the realization of common goals can be facilitated: the reduction of major accident hazard risks and consistent performance of assets. For a sector on which many emergent national decarbonization policies are reliant upon to deliver "net zero", an all-encompassing risk based safety and AIM approach offers asset owners and operators in the beginnings of the hydrogen economy boom a means by which to deliver business objectives profitably without major incident.

625565 More Haste, Less Speed: Proportionate Supply Chain Risk Management for the Accelerating Hydrogen Economy

Abstract:

The long awaited acceleration of the hydrogen economy is dawning. It is viewed as a necessity in the race to decarbonize industry, heat and transport, amongst other key sectors. Existing hydrogen economy supply chains serving the current markets face the challenge of ramping up to meet future demand. Aside from hydrogen growth, many organizations are increasingly concerned about supply chain risk, not only the traditional supplier issues of late delivery and poor quality but also other risks with security, environmental, social and ethical impacts. Furthermore, hydrogen as an energy vector requires significant safety management to mitigate its potential to cause major accidents. The confluence of increasing demand, major accident hazards associated with hydrogen and concerns of supply chain risks present a significant challenge for those organizations in the hydrogen supply chain. When viewed through the lens of rapid demand acceleration, supply chain risks are magnified by the potential for short-cuts and risk taking. Such risks are exacerbated by the potential set-back in the global decarbonization goal, should a major accident be realized. Formal Supply Chain Risk Management (SCRM) techniques adopting a risk based approach are explored to help mitigate supply chain risks in a proportionate way, reducing supply chain vulnerability, ensuring continuity if risks occur and increasing value within the supply chain through improved performance. Crucial attributes of successful SCRM are considered to be: application of good governance, early supplier engagement, proportionate assurance levels based on risk, supply chain mapping, monitoring of risk events/risk event pre-cursors and risk register maintenance.

639742 Leading and Empowering Global Teams during a Pandemic

Abstract:

After almost two years of a global pandemic, businesses and life in general have had to adapt to a new normal. Most of us found ways to effectively perform our jobs from a forced work-from-home environment, but did we do it right? Did we lose something more than the casual coffee chat with colleagues at the office? Did we lead our teams in an effective way? In this talk, we will discuss and reflect on a few examples from ExxonMobil.

638117 Navigating the COVID-19 Pandemic at Michigan Technological University

Abstract:

Very few could have imagined that in early 2020 the novel coronavirus would upend the normal life (including academia) in the US. During the 2020 spring break, it was decided that students would not return to campus, and that all classes would be taught remotely for the remainder of the spring term. The campus was shut down in the third week of March 2020, with no access to offices and other campus facilities. Most research laboratories were also shuttered. This shutdown presented several challenges both at the administration level and at the academic unit level. Very few faculty members had the knowledge and experience of teaching an on-line course. With no historical precedent to guide a situation like this, the academic community had to adjust and adapt quickly. A decision was made to offer classes in blended mode (i.e., in-person + remote) for the 2020-2021 academic year. The efforts during summer 2020 included (i) training faculty for online teaching, (ii) scheduling courses in blended mode while following the CDC safety protocols, and (iii) making special provisions for teaching the Unit Operations Laboratory courses while still preserving hands-on experiential learning. Both the UO Laboratory and capstone senior design courses also overcame the challenge of holding help/review and presentation sessions for student teams. The efforts across the curriculum were made to accommodate students who required isolation because of symptoms or possible exposure to COVID-19 and to provide remote access to design software. Overall, it has been nothing less than a Herculean effort to pull this off successfully. This presentation will share the challenges and operational details and provide insights into those changes that will become permanent enhancements to the program.

636697 Managing Teams Effectively during a Pandemic Required – Empathy, Connection, Agility, Prioritization and Support

Abstract:

COVID-19 Global Pandemic has catalyzed growth in leadership behaviors and effectiveness. Progressing on innovation goals while prioritizing employee and customer experiences has required new approaches and solutions. We will discuss examples from Dow Industrial Solutions R&D that focused on Empathy, Connection, Agility, Prioritization and Support.

636972 Lessons learned from field testing a new technology during the pandemic

Abstract:

During 2020 and 2021, NICE America conducted field tests of a new hydrogen refueling technology during two separate campaigns, each lasting over three months. During normal times, this endeavor would have required coordination of a multinational team spread across three time zones, safe execution of an ambitious technical validation program, and management of expectations for internal and external stakeholders (including potential customers). The pandemic introduced additional challenges including compliance with evolving public health regulations, constraints on travel and scheduling, and supply chain stress associated with economic consequences of pandemic response. This presentation will describe the actual campaigns, review lessons learned from adapting to the evolving situations, and highlight effective practices that can be generalized for field testing under less extreme circumstances.

AUTHOR INDEX

Adesope, Olusola 5
Aghalayam, Preeti 1
Agrawal, Pradeep K. 39
Aui, Alvina 35
Azarin, Samira 7
Bademi, Apoorva 34
Barr, Christopher 7
Bazant, Martin Z. 4
Brennan, Janie 7
Campbell, Zachary S. 13
Cao, Shijie 19
Carey, Thomas R. 16
Carter, Tracy 7
Chen, Joyce 23
Cheung, Timothy 20
Ciston, Shannon 28
Collins, Jeanna 39
Cooper, Matthew 13
Cox, Brian 20
Daniel, Susan 24
Donders, Eric 20
Duggal, Rajat 40
Dutta, Prashanta 5
Ellor, Gareth 36, 37
English, Niall 6
Falzarano, Darryl 20
Farlessyost, William 34
Gartner, Jacqueline Burgher 5
Goluch, Edgar D. 25
Gor, Gennady 12
Gu, Zongyu 4
Guzman, Javier 38
Heller, Daniel 26
Huayta, Javier 13
Hubbell, Jeffrey A. 19
Isaacs-Bernal, Daniela 20
Islam, Mohammad Aminul 15
Kaiphanliam, Kitana 5
Karimi, Iftekhar 32
Karlsson, Amy J. 7
Katragadda, Apoorva 32
Khan, Aminul Islam 5
Koretsky, Milo 9
Kozminsky, Molly 16
Ku, Anthony 41
Landherr, Lucas 8
Lepek, Daniel 14
Liberatore, Matthew 14
Luks, Christi Patton 10
Martin, Irene Mas 35
Monty-Bromer, Chelsea 27
Nosi, Ursula 20
Oje, Olufunso 5
Preusser, Deja 9
Raczy, Michal 19
Reynolds, Olivia 5
Rickard, Kurt A. 39
Rogers, Tony N. 39
Schulte, Lisa 35
Shoichet, Molly S. 20
Singh, Shweta 34
Slaughter, Kai 20
Sohn, Lydia L. 16
Stevens, Miriam 34
Swartz, Melody 19
Thiessen, David B. 5
Vaidyanathan, Kartic 1
Van Wie, Bernard 5
Versypt, Ashlee Ford 15
Volpatti, Lisa 19
Wallace, Rachel 19
Wang, Ruyi 19
Wang, Xiaonan 32
Wilson, Sarah 7
Woldemariam, Tekeleselassie 20
Wong, Amy 20
Wright, Mark Mba 35
Xue, Chang 20
Yaari, Zvi 26

AIChE
120 Wall Street, FL 23
New York, NY 10005-4020

CURRAN ASSOCIATES INC.
proceedings.com

ISBN 978-1-7138-5684-9

Pharmaceutical Discovery, Development and Manufacturing Forum 2022

Held at the 2022 AIChE Annual Meeting

Phoenix, Arizona, USA
13-18 November 2022

AIChE
The Global Home of Chemical Engineers